Essential Treatments in Cardiovascular Chinese Medicine 1: Hyperlipidemia

Essential Treatments in Cardiovascular Chinese Medicine 1: Hyperlipidemia

Dr. Anika Niambi Al-Shura, BSc., MSOM, Ph.D
Continuing Education Instructor
Niambi Wellness
Tampa, FL

Medical Illustrator: Samar Sobhy

ELSEVIER

AMSTERDAM • BOSTON • HEIDELBERG • LONDON
NEW YORK • OXFORD • PARIS • SAN DIEGO
SAN FRANCISCO • SINGAPORE • SYDNEY • TOKYO

Academic Press is an imprint of Elsevier

Academic Press is an imprint of Elsevier
32 Jamestown Road, London NW1 7BY, UK
The Boulevard, Langford Lane, Kidlington, Oxford, OX5 1GB, UK
Radarweg 29, PO Box 211, 1000 AE Amsterdam, The Netherlands
225 Wyman Street, Waltham, MA 02451, USA
525 B Street, Suite 1900, San Diego, CA 92101-4495, USA

Notices
Knowledge and best practice in this field are constantly changing. As new research and experience broaden our understanding, changes in research methods, professional practices, or medical treatment may become necessary. Practitioners and researchers must always rely on their own experience and knowledge in evaluating and using any information, methods, compounds, or experiments described herein. In using such information or methods they should be mindful of their own safety and the safety of others, including parties for whom they have a professional responsibility.

To the fullest extent of the law, neither the Publisher nor the authors, contributors, or editors, assume any liability for any injury and/or damage to persons or property as a matter of products liability, negligence or otherwise, or from any use or operation of any methods, products, instructions, or ideas contained in the material herein.

British Library Cataloguing-in-Publication Data
A catalogue record for this book is available from the British Library

Library of Congress Cataloging-in-Publication Data
A catalog record for this book is available from the Library of Congress

ISBN: 978-0-12-800119-6

For information on all Academic Press publications
visit our website at **store.elsevier.com**

This book has been manufactured using Print On Demand technology. Each copy is produced to order and is limited to black ink. The online version of this book will show color figures where appropriate.

DEDICATION

The energy and effort behind the research and writing of this textbook is dedicated to my son, Khaleel Shakeer Ryland. May this inspire and guide you through your journey in your medical studies, career, and life.

ACKNOWLEDGMENTS

This is a special acknowledgement to my seven-year medical students at Tianjin Medical University (2012—2013) who served as cardiovascular research assistants. May your future medical careers be successful.

An Qi He
Bin Lin Da
Han Jiang
Chen Hua
Jia Ying Luo
Jun Zhang
Lin Lin
Ming Lu
Nang Zhang
Ping Tang
Hu Si Le
Zhao Tian Man
Wen Xing Ning
Xing Wen Zhao
Tang Ying Mei
Li Ying Ying
Xiong Yong Qin
Ding Yu
Li Yan Jun

CONTENTS

NIAMBIWELLNESS
INTEGRATIVE CARDIOVASCULAR CHINESE MEDICINE

The companion course which is required for study with this textbook edition can be found on the Elsevier website and at www.niambiwellness.com.

APPROVING AGENCIES

PROFESSIONAL
NCCAOM
DEVELOPMENT ACTIVITY

The course with this textbook is entitled, Integrative Hyperlipidemia Treatments in Cardiology.

This course is approved by the National Certification Commission for Acupuncture and Oriental Medicine (NCCAOM), and is listed as course #1053-005 for 5 PDA points.

This course is approved by the Florida State Board of Acupuncture, and is listed as course #20-334885 for 5 CEU credits.

COURSE DESCRIPTION

This course covers the basic Western medicine and Traditional Chinese Medicine sciences of blood lipids, clinical significance, differentiations, diagnoses, treatment principles and methods.

COURSE OBJECTIVES

- Describe the basic etiology of lipoprotein disorders, including genetic disorders.

- Describe the clinical presentations of lipoprotein disorders as a sign of cardiovascular diseases.
- Describe Western medicine and Traditional Chinese medicine prescriptions for lipoprotein disorders.
- Describe the suggested medicated diet for patients on a treatment course.

CHAPTER *1*

Etiology of Lipoprotein Disorders in Chinese Medicine and Western Medicine

CHAPTER OBJECTIVES

After studying this chapter, you should be able to:

1. Explain the Chinese medicine classification of lipoprotein disorders.
2. Explain the Western medicine perspective on lipoprotein disorders.
3. Describe the genetic subtypes of familial hypercholesterolemia (FH).
4. Describe the genetic apolipoprotein isoform mutations and their role in lipoprotein disorders and cardiovascular diseases.

1.1 PART 1: INTEGRATIVE PERSPECTIVE

1.1.1 Lipoprotein Disorders in Chinese Medicine

Category	Sub categories	Classification
–Tan zhuo –Xue yu –Xuan yuan	Headache, strokePalpitation, obstruction, Vertigo	–Blood stasis –Turbid phlegm obstruction.
Lipoprotein disorders may be generalized as a deficiency of the liver, spleen and kidney. Liver overaction upon the spleen results in chronic digestive problems.		

Copyright © 2014 Anika Niambi Al-Shura. Published by Elsevier Inc. All rights reserved.

1.1.2 Lipoprotein Disorders in Western Medicine

The lipid transport system moves lipids from the hepato-biliary and gastro-intestinal system to areas of the body where it is needed to produce certain hormones.

Essential Treatments in Cardiovascular Chinese Medicine 1: Hyperlipidemia.
DOI: http://dx.doi.org/10.1016/B978-0-12-800119-6.00001-2

1. What factors commonly cause lipoprotein disorders?

_____

Log on and complete the companion course at www.niambiwellness.com for the answers.

1.1.2.1 Hepatic and Intestinal Pathways

Cholesterol (VLDL) is sent from the liver into the small intestine for chylomicron processing.

2. What is the next step?

_____

Log on and complete the companion course at www.niambiwellness.com for the answers.

1.1.2.2 Renal and Hepatic Disorders

Patients with chronic renal failure often have elevated triglyceride levels with a low HDL.

3. What problems to patients with glomerular nephritis have?

_____

Log on and complete the companion course at www.niambiwellness.com for the answers.

1.2 PART 2: GENETIC DETERMINANTS

1.2.1 Familial Hypercholesterolemia

These genetic determinants include a predisposition for accumulation of triglycerides, LDL, and cholesterol at specific sites on the body.

4. What three types of organ disorder do they commonly cause?

_____

_____

_____

Log on and complete the companion course at www.niambiwellness.com for the answers.

Type I Familial hyperchylomicronemia
–Lipoprotein lipase (LPL) deficiencies –Elevations in triglycerides –Reduced HDL levels.
Type II Familial hypercholesterolemia
–PCSK9 mutations with elevated cholesterol levels –LDL-R gene defect with elevated LDL levels –Apo B gene defects and high elevations of LDL –Linked to coronary artery disease –Prevalence in the U.S. African Americans
Type III Familial hyperlipoproteinemia
–Apo B gene defects –PCSK9 mutations, –LDLRAP1 mutations –ABCG 8 mutations –Excess liver VLDL makes excess cholesterol –Xanthomas –High cholesterol –Metabolic syndrome
Type V Familial hyperlipoproteinemia
–Combination of type I and IV –Splenomegaly –Xanthomas

1.2.2 Apolipoprotein B (Apo B) Isoforms B100 and B48

Apo B100	–Elevated levels due to coronary artery disease due to Staphlococcus aureas, leaving cholesterol streaks.
Apo B48	–Excess in chylomicrons raises LDL levels and complicates Diabetes Mellitus treatment.
Apolipoprotein E (Apo E)	E2: slow fat metabolism, increased risk of coronary artery disease
	E3: neutral allele
	E4: atherosclerosis, sleep apnea, ischemic cerebrovascular disease

What are three significant factors related to Apo B isoforms?

5. _____

6. _____

7. _____

Log on and complete the companion course at www.niambiwellness.com for the answers.

NOTES

Log on at www.niambiwellness.com to access the companion course and quiz for Module 1.

Clinical Presentations

CHAPTER 2

Western Medicine Examination and Diagnosis of Lipid Disorders

CHAPTER OBJECTIVES

After studying this chapter, you should be able to:

1. Explain the steps during the physical examination.
2. Explain what possible symptoms may be reported by the patient.
3. List the common tests done in the lipid panel and the values.

2.1 PART 1: EXAMINATION

2.1.1 Visual Observation

Search for lipid accumulation on the face and areas of the body.

1. Where should you pay the most attention?

_____

Log on and complete the companion course at www.niambiwellness.com for the answers.

2.1.2 Inspection

Area	Possible symptoms
Head	Dizziness, distended feeling and heavy feeling
Eyes	Xanthomas
Mouth	Sticky taste sensation and greasy tongue coating
Emotions	Moody and restlessness
Limbs	Pain, numbness and tingling
Others	Hormonal problems and weight gain

Copyright © 2014 Anika Niambi Al-Shura. Published by Elsevier Inc. All rights reserved.

A. Which areas should be considered during the patient narrative?
B. Which areas should be considered during physical examination?

Essential Treatments in Cardiovascular Chinese Medicine 1: Hyperlipidemia.
DOI: http://dx.doi.org/10.1016/B978-0-12-800119-6.00002-4
© 2014 Elsevier Inc. All rights reserved.

2.1.3 Physical Examination
Measure the waist circumference and weight.

2. What body shape is considered an example of cardiovascular problems?

_____

Log on and complete the companion course at www.niambiwellness. com for the answers.

2.1.4 Blood Pressure
Elevated levels can indicate possible chronic hypertension.

3. What can patients do at home to manage symptoms?

_____

Log on and complete the companion course at www.niambiwellness. com for the answers.

2.2 PART 2: DIAGNOSIS

Lipid type	mg/dL	mmol/L	Interpretation
Total cholesterol (TC)	<200	<11.1	Normal
	200–239	11.1–13.3	Borderline
	>240	>13.3	High
LDL	<100	<5.5	Normal
	100–129	5.55–7.15	Good
	130–159	7.15–8.82	Borderline
	160–189	8.82–10.5	High
	>190	>10.5	High risk
HDL	<40	<2.21	High risk
	41–59	2.21–3.27	Borderline
	60	3.27	Good

The lipid panel provides information on TC (total cholesterol), LDL (low-density lipoprotein), and HDL (high-density lipoprotein).

4. What do these tests help with?

_____

Log on and complete the companion course at www.niambiwellness. com for the answers.

NOTES

Log on at www.niambiwellness.com to access the companion course and quiz for Module 2.

For chapter 2 tutorial log on at www.niambiwellness.com to access the companion course for module 2.

Prescriptions

CHAPTER 3

Stages of Lipoprotein Disorders and Syndrome Differentiation in TCM

CHAPTER OBJECTIVES

After studying this chapter, you should be able to:

1. Explain the clinical presentations for stage 1.
2. Explain the clinical presentations for stage 2.
3. Explain the clinical presentations for stage 3.

In Chinese medicine, acquired lipoprotein disorders are the result of liver, spleen, and kidney deficiencies and can develop in stages.

3.1 STAGE 1: DEFICIENCY OF LIVER AND KIDNEY YIN

Clinical Presentations	Blurred vision, irritability, dizziness, soreness and weakness of the limbs, low back knees, vertigo, tinnitus, insomnia, forgetfulness, dream filled sleep and poor appetite
Tongue	Scanty fur
Pulse	Thready and tight

Deficient kidney yin fails to nourish the liver yin.

1. What could liver yang hyperactivity and qi stagnation contribute to or directly cause?

_____

Log on and complete the companion course at www.niambiwellness.com for the answers.

Essential Treatments in Cardiovascular Chinese Medicine 1: Hyperlipidemia.
DOI: http://dx.doi.org/10.1016/B978-0-12-800119-6.00003-6

3.2 STAGE 2: DAMPNESS RETENTION LEADING TO DAMP HEAT

Clinical Presentations	Irritability, depression, headache or head distention, bitter and dry mouth, poor appetite, heat sensation in the chest, heavy body feeling, lower abdominal pain, low back pain, false sensation of urinary urgency with dark scanty urine and dry constipation
Tongue	Pale with yellow greasy coat
Pulse	Rolling or smooth

Spleen and kidney yang deficiency leads to excess fluid retention in the body. Fluid stagnates in the interstitium, insulating, accumulating precipitation, and warmed by body heat.

2. What is the unfavorable result?

_____

Log on and complete the companion course at www.niambiwellness. com for the answers.

3.3 STAGE 3: PHLEGM STAGNATION AND STASIS

Clinical Presentations	Obesity, dizziness, chest fullness, abdominal distension, heavy head, heavy body sensation, phlegm accumulation in the sinuses prompting drawing and spitting, tinnitus, bitter mouth taste with a sticky feeling in the mouth with need to drink more water and poor appetite

Spleen and kidney yang deficiency leads to excess fluid retention and lin syndrome.

3. What does the dampness lead to?

_____

Log on and complete the companion course at www.niambiwellness. com for the answers.

NOTES

For chapter 3 tutorial log on at www.niambiwellness.com to access the companion course for module 2.

Module Review Questions

1. Explain the steps during the physical examination.
2. Explain what possible symptoms may be reported by the patient.
3. List the common tests done in the lipid panel and the values.
4. Explain the clinical presentations for stage 1.
5. Explain the clinical presentations for stage 2.
6. Explain the clinical presentations for stage 3.

Log on at www.niambiwellness.com to access the companion course and quiz for Module 2.

The Medicated Diet

CHAPTER 4

Actions of Lipoprotein Prescriptions in Chinese Medicine and Western Medicine

CHAPTER OBJECTIVES

After studying this chapter, you should be able to:

1. Explain the common action of Western medicine drugs for lipoprotein disorders.
2. Discuss a primary treatment principle based on differentiations in Chinese medicine.
3. Compare and contrast the medicine actions between Chinese medicine and Western medicine.

4.1 PART 1: WESTERN MEDICINES USED FOR LIPOPROTEIN DISORDERS

Medicine	Action
Ezetimibe	Lowers cholesterol absorption in the small intestine.
Statins	Inhibits HMG-CoA reductase enzyme for cholesterol reduction.
Bile acid sequestrants	Prevents fat absorption in the intestines.
Niacin	Helps break down fat tissue.
Fibrate	Combines with statins to treat high cholesterol.

Copyright © 2014 Anika Niambi Al-Shura. Published by Elsevier Inc. All rights reserved.

1. What are the side effects of ezetimibe?

   _____

2. What are the side effects of statins?

   _____

3. What are the side effects of bile acid sequestrants?

   _____

Essential Treatments in Cardiovascular Chinese Medicine 1: Hyperlipidemia.
DOI: http://dx.doi.org/10.1016/B978-0-12-800119-6.00004-8

4. What are the side effects of niacin fibrate?

_____

Log on and complete the companion course at www.niambiwellness.com for the answers.

4.2 PART 2: CHINESE MEDICINES USED FOR LIPOPROTEIN DISORDERS

Chai hu shu gan san	
Reduces liver hyperactivity, relieves body distention and nourishes kidney yin and blood.	
Buplerum	Soothes the liver and activates qi
Nutgrass flatsedge rhizome	Soothes the liver and activates qi
Szechuan lovage rhizome	Circulates blood and activates qi
Tangerine peel	Regulates qi
Immature bitter orange	Regulates qi
Peony root	Nourishes blood and tonifies liver
Licorice root	Nourishes blood and tonifies liver

5. What are the contraindications of this formula?

_____

Log on and complete the companion course at www.niambiwellness.com for the answers.

Weicao (capsule) tang	
Reduces blood lipids and increases lipid metabolism.	
Clematis root	Dredges collateral and relieves pain
Cassia seed	Clears liver and nourishes kidneys
Motherwort	Activates blood
Lysimachia	Clears heat and removes toxicity
Crude rhubarb	Dredges the bowels
Peony root	Nourishes blood and tonifies liver
Licorice root	Nourishes blood and tonifies liver

6. What are the contraindications of this formula?

_____

Log on and complete the companion course at www.niambiwellness.com for the answers.

Eliminate Blood Lipid Formula	
Reduces blood lipids for weight loss.	
Glossy privet fruit	Replenishes yin and reduces dampness
Fleece flower root	Dredges bowels and reduces dampness
Siberian solomon seal rhizome	Tonifies spleen and reduces dampness
Fragrant Solomon seal rhizome	Nourishes yin and reduces dampness
Dan shen root	Calms the mind and clear heat
Chinese angelica	Deficiency of heart and spleen
Spine date seed	Calms the mind and nourishes heart

7. What are the contraindications of this formula?

_____

Log on and complete the companion course at www.niambiwellness. com for the answers.

Shen qi wan	
Helps to tonify and strengthen the kidneys.	
Buplerum	Soothes the liver and activates qi
Nutgrass flatsedge rhizome	Soothes the liver and activates qi
Szechuan lovage rhizome	Circulates blood and activates qi
Tangerine peel	Regulates qi
Immature bitter orange	Regulates qi
Peony root	Nourishes blood and tonifies liver
Licorice root	Nourishes blood and tonifies liver

8. What are the contraindications of this formula?

_____

Log on and complete the companion course at www.niambiwellness. com for the answers.

Reduce Blood Lipid Formula	
Reduces LDL levels.	
Glossy privet fruit	Replenishes yin and reduces dampness
Wolfberry fruit	Nourishes kidneys and liver and reduces lipids

9. What are the contraindications of this formula?

_____

Log on and complete the companion course at www.niambiwellness. com for the answers.

4.3 PART 3 MEDICATED DIET PLAN

In the companion course, general recipes are described as a base for including medicinal herbs later.

Breakfast savory: hai zao, kun bu, tian ma and jue ming zi Breakfast sweet: shanza and gou qi zi
Lunch savory: hai zao, kun bu, tian ma, yu jin, suan, he shou wu Lunch sweet: shanza and gou qi zi
Dinner savory: hai zao, kun bu, tian ma, yu jin, suan and he shou wu Dinner sweet: shanza and gou qi zi, tian ma, yu jin, suan, he shou wu

10. Which Chinese medicine formula(s) in section 4.2 may be appropriate with this medicated diet recipe plan?

_____

Log on and complete the companion course at www.niambiwellness. com for the answers.

Breakfast savory: hai zao and ju hua Breakfast sweet: shanza, gou qi zi, da zao and dang shen
Lunch savory: hai zao and ren shen Lunch sweet: shanza, gou qi zi, da zao and dang shen
Dinner savory: hai zao, long gu, mu li and tian ma Dinner sweet: shanza, gou qi zi, da zao and dan shen

11. Which Chinese medicine formula(s) from section 4.2 may be appropriate with this medicated diet recipe plan?

_____

Log on and complete the companion course at www.niambiwellness. com for the answers.

NOTES

Log on at www.niambiwellness.com to access the companion course
and quiz for Module 3.

For chapter 4 tutorial log on at www.niambiwellness.com to access
the companion course for module 4.

Module Review Questions

1. What are common actions of Western medicine drug for lipoprotein
   disorders?
2. What may be the differentiations with each of the formulas for lipoprotein disorders?
3. Which formulas may be appropriate for which herb combinations in
   the medicinal diet plans?

Log on at www.niambiwellness.com to access the companion course
and quiz for Module 3.

This also concludes the Integrative Anatomy and Pathophysiology in TCM Cardiology course. It is strongly suggested that you log onto the courses at the companion websites to review the course modules. Next, submit course documents and complete the final exam.

Upon passing the exam, you will receive completion certificates that include your name and practice license number, along with the specific number of credit hours awarded for this course. Electronic transmission of CEU and PDA credits will be sent to NCCAOM and your state medical board.

www.ingramcontent.com/pod-product-compliance
Lightning Source LLC
Chambersburg PA
CBHW072157020426
42334CB00018B/2053